# poems for paco

kovida

# poems for paco

Mayajala Books

mayajala

First Published in 2001
By Mayajala Books,
30, Chantry Road,
Moseley,
Birmingham,
B13 8DH.



ISBN 0-9539857-0-9

Printed and bound in Malta by Interprint Limited

# Acknowledgements

I'd like to thank the following people for their help in the production of this book of poems –

Dhammarati for the front cover and advice.
Simon Perry for the layout.
Dhivati for proof reading.
Padmavajri and Dharmashura at
Windhorse Publications for all their assistance.
Francisco Domínguez Montero for the translation of 'Motorway'.
Urgyen Sangharakshita for having the patience to read through the whole thing and to very kindly make helpful and constructive comments.
Paco for allowing me to publish our love life so fully.

# Preface

This collection of Poems deals with my response to a love affair with a Spanish boy called Paco, that began some time in August 1995 and ended in New York in 1997. However, the friendship upon which it was based has grown and strengthened over the years. Of course it was not possible for me to leave off my attachments immediately and when Paco got a new boyfriend later in 1997 this again proved a source of pain and consequently more poetry.

I have written mainly from my own sense of pain and joy but have tried from time to time to write placing myself in his shoes...

The poems that appear here have been selected from a far greater number on the same theme and tend to begin in late 1996. I have arranged them chronologically as this seems to echo the ebb and flow of our relationship, although some of the dates are conjectural.

One other thing that may be pertinent is that we were both involved in the practice of Buddhism, although over the last few years Paco has changed the focus of his interest and is now no longer involved.

Kovida.

13.09.00

# Contents

# Paco 1

I need to have you need me to maintain
This fragile sense of self – this ego 'I'
For through your love of me, I can attain
A state of bliss where all my sorrows die.
And yet, I know too well this cannot last,
For beauty such as yours is too divine
To be defined by my conditioned past,
Or trapped within my selfish love's confine.
For I have grown dependent on your love,
Show weakness, and am vulnerable to loss,
And seek your smiled approval for each move
And crucify myself upon your cross.
  I need to have you need me to maintain
  This thing called love, this ... self-indulgent pain.

24.08.96

# Paco 2

You came into my life like one divine,
Your dark mysterious beauty rich and pure,
Your graceful, fluid movements, so sublime,
Your confidence, your sense of self – secure.
Your mind was razor-sharp, your eyes were bright,
Your laugh of boyish humour rang aloud,
Your beauty filled my life with pure delight,
To know you, and to love you, made me proud.
If I had been content to let you be,
To love you, then release you from my grip,
Our love would still be fresh, our hearts be free,
Our soaring wings, far from the hunter's clip.
  But Jealousy, Possession, called you mine,
  And thus destroyed what once was so divine.

24.08.96

# Stillness

There is a vital stillness in your being,
A calm and self-contentment unbeknown,
And foreign to my normal way of seeing
This strange and hostile world through which I'm blown.
To you, being 'Gay' is simple as the sun,
Which shines without distinction on all life.
To me, being 'Gay' is something never done,
A war, an endless struggle, constant strife.
You like all men, while I like only boys:
You do not judge, but smiling, welcome all,
While I restrict and censure all my joys,
Thus guilty agitation keeps me thrall.
  Your stillness and contentment hold a key
  To traits I must relinquish to be free.

24.08.96

# Denied

*That you existed and I did not run to hold you is my eternal shame - forgive me.*

After I had held you,
I had to set you free.
Trapped, you were my albatross,
And thus the curse of me.

Free, you are the gentle wind,
Blowing through the trees.
Free, you are my muse's breath,
Like summer's healing breeze.

All that I do is touched by you,
Angel in human form.
You came to me, to set me free,
Spirit, on earth, reborn.

The love I denied
Was frozen inside,
And my grip of steel,
Could never reveal
How your gentle spirit died.

But, when I let go,
Your joy could flow,
From out of your open heart.
And when I released you,
Our friendship increased too,
And now we are never apart.

01.07.97

## Innocent

All spruced up like a sweet young child,
Sparkling, new, and fresh;
Your skin smooth,
Your hair trimmed,
Your clothes all washed and pressed.

Off, like a lamb to the slaughter,
Innocent and wide-eyed,
To the arms of your (so cold) lover,
Who must not be denied.

Oh how your young sleeve proudly wears
The love that pulsed in your veins.
Oh how the world must know and see
The love that your heart contains.

Oh how you desperately, desperately,
Wanted to be loved.
Oh how he was right for you,
Or so your keen voice cried.

Oh how you made the effort,
Oh how you put in the hours,
Oh how you did the right, good thing,
To make his cold heart flower.

And all this time,
My own heart bleeds,
For my  unfulfilled raw needs.

"It should have been me!"
Blares out its creed,
from the thankless, stark, CD.

'Funny how love can be.

Love.

Funny how love can be!'

01.07.97

## New lover

You told me of your lover, smilingly,
As if I could, with open heart, rejoice.
(If you thought that, you've over-valued me,
Saw Virtue in a heart that's laced with Vice.)

I cannot wish you happiness or joy,
Call blessing on you, while I suffer pain;
You were my love, my life, my angel boy,
And now you're his, while he's become my bane.

I'd rather not have intercourse with you,
Meet up, or keep this false 'delight' alive.
For all I do in being friends with you
Merely seems to help your lover thrive.

One day, when I am old, and you are grey,
Our love a long forgotten brief affair,
I'll look back fondly, smile at this sad day,
And hope I'd acted honourably and fair.

Today, however, feelings run amok,
And choke the thoughts of pleasantness and joy,
They cast all light behind Hate's dark red smoke,
And thus deny the love we did enjoy.

So go to him, make love, pour joy on him,
And leave me in my squalor and my pain,
I'll struggle with my anger and, in time,
Will see your virtues, love you once again.

01.07.97

## Loving

*"Loving is more the essence of friendship than being loved" – Aristotle.*

I love you, though you love me not,
For all the things you do;
For being young and beautiful
So faithful and so true.

I love you for your frailty,
Your innocence and youth,
I love you for your intellect,
Which bares the light of truth.

I love you for your beauty,
So dark, so rare, so fine,
I love you for your humour,
And your wisdom so divine.

I love you for your care for me
Your genuine concern.
I love you for your skilfulness,
Unwavering and stern.

I love you when you challenge me,
And force me to revise
The view I have about myself,
To see though altered eyes.

I love you for the high ideals
You bring before my mind.
I love your for your courtesy,
So gentle and refined.

But no matter how I love you,
I think that in the end
I really love you most of all
Because you are my friend.

01.07.97

# Your Circumcision

*(A phantasy of being you, lying there in hospital.)*

I lie, face up, upon my pain-filled bed.
My penis cut and bloodied by their knife.
Tears well up, and darkness floods my head.
Who knows my anguish? Who can feel my pain?
Who sympathise? Who comprehend my life?
And who can give me love, or bring me gain?

The room is silent now, the ward serene;
The visitors departed to their homes,
Their laughter and their present joy obscene,
Compared to what I feel in this dark place.
Thoughts take me on a journey. Body Groans.
And Anguish writes His name across my face.

"All things must pass. All pain will fade away."
These words bring comfort, and their own despair,
How often does this Wisdom slip away
And leave me in the rawness of my hurt,
Their comfort gone, dispelled like rancid air,
And Hope dashed down and trampled in the dirt.

Bad timing, long delayed, postponed and passed,
This dreaded operation now has come,
And pain has maimed my joyful youth at last.
Yet, this is but a torment of the flesh,
And shouts its hurt, while mental pain is dumb,
Too subtle for the senses to enmesh.

We are such fragile creatures, you and I,
A sack of skin, a skeleton of bone,
Our life is but a mirage, then we die,
And wonder what the purpose of it was.
I want to live my life for me alone!
I want to be myself, have my own cause!

No more by others' judgements be confined,
No more by others' dictates taught to act.
My own man now, the King of my own mind,
I'll go my *own* way, live the life I choose,
I'll cease to be with endless conflicts racked,
I'll take my chance, what else have I to lose?

The pain returns. I vomit on the bed.
I press the buzzer. Nurse comes running in.
Administers some drug that dulls my head.
The pain has gone. I drift on clouds of white.
My parents bless me. Father talks of sin.
My sister smiles, and fills me with delight.

Huh! How quickly our resolves are washed away,
As we forget that no conditions last,
Our goals – 'Today' dissolve in – 'Yesterday'.
Our optimistic vision fades and dies,
And thoughts we had now drift into the past,
To cover up our failings with fresh lies.

Sweet Hospital! (How foolish can I be?
What's sweet about a place that brings you pain?)
Oh would that I could set my spirit free
And soar above all petty doubt and care,
Live out my life! My own true goal attain,
And end all self-despising and despair.

I value love, nay, need it, want it, too.
And Friendship is a key to all my life,
I love Communication: pure and true,
And learn my thinking speaking thoughts to you.
But what to do, and why, still brings me strife,
My existential doubt rears up anew.

I lie, face up, upon my pain-filled bed,
My penis cut and bloodied by the knife
Dreams of sadness swirl around my head,
And I, though much provoked, have no solution,
Know nothing of the way to live my life,
And see, for me, no simple resolution.

I'm young and want to taste the joys of life,
Be wild, and dance, and sing, and laugh, and play,
Be free from guilt, and this internal strife,
That rives my mind with conflict and despair.
I want to be immoral: camp and Gay!
Yet follow moral teachings, true and fair.

My mind's so sharp it cuts me – like their knife –
Serrates me, till I drop from excess thought,
And fall a bloody victim to dull Life,
Lived by one who lives below his best.
Where can I go? What do? – You answer not!
Dear Friend, I need you now, to pass this test.

My mind flits in and out of consciousness.
Dreams drift over eyes, then skies are blue!
My thought returns. I gain my seriousness,
Then drift some more, fresh pastures to explore.
And in this state, I wonder what is True,
The dreams I have, or wounds that make me sore?

There is no easy answer to my plight,
The curse of Mindfulness undoes my life
Consumes me in its all-embracing light
And leaves me shaking, cold and wet with sweat.
Who can, when once Aware, and thoughts are rife,
Deny what they have known, and thus forget?

I lie, face up, upon my pain filled bed.
My penis cut and bloodied by their knife,
Tears well up, and darkness floods my head.
Who knows my anguish? Who can feel my pain?
Who sympathise? Who comprehend my life?
Oh who can bring me Ignorance again?

09.10.97

# Miss you

The pain of my longing
Was harder than hunger.
The pain of our parting
Was smarting and smarting.
Oh darling, I miss you.
Oh god, how I miss you!
I want to be near you,
To hold you, to kiss you.
Oh god, how I miss you!
My darling. My love.

21.12.97

# Love's breath

Do not chastise me for the love I feel,
To you it is a pain, to me it's real.
I beg you, for my sake, please thus endure,
For what I feel for you is true and pure.
I know that you abhor these joys I steal,
And think me weak, for you are strong and sure,
And rich in your affections, where I'm poor,
But know, my friend, this world is not ideal.
You too, will one day suffer my disease,
Will find a lover, cling to him in hope,
And gain all strength and purpose from his breath.
And then, like me, you'll taste Love's fickle breeze,
Which blows where'er it will and knows no scope,
And wafting on, leaves sorrow, pain, and death.

23.03.98

## Pain Two

Think not that I would cause you further pain,
Or be the cause of suffering or sorrow,
I know too well the tensions and the strain
That being in love can bring with each new morrow.

I too have loved – am still in love with you –
But you have placed affection in another,
So I must change my love and be a brother,
And free you to enjoy the love you do.

24.03.98

# Friendship

I never meant to hurt you, or cause harm,
Or leave you, as you sailed through Love's dark storm.
I loved you with a love I thought you knew,
And always hoped that I'd be true to you.
Yet I confess that Jealousy was rife,
And I in cruelty, bitchiness, and hurt,
Condemned your love, and worked to cause you strife,
By mocking where you chose to place your heart.
But now I know, you love the one you do,
And need his love to make your world complete.
How foolish then to think I could compete
Or even come between his love and you.
    So now I hide away and feel remorse,
    And understand that Love's not won by force.

25.03.98

# On our inability to relinquish self-consciousness

This human Life's a struggle without end
When we commit ourselves to higher goals
For powers with which we must at first contend,
Force us to examine all our roles.

We cannot kill self-consciousness and flee;
Not choose the skilful path in mindfulness;
Nor turn blind eyes to how our lives must be,
Remorse ensures a pristine Consciousness.

For, burning like an everlasting flame,
Self-consciousness, awareness, leads us on,
Until all sense of self, and hope of gain,
Vanish, and our ego, 'I', is gone!

26.03.98

# Your Dear Friend

The ease with which you open up your heart
Has always been a source of joy to me
No 'self-important' vision to impart,
But Truth alone, self-knowledge given free.
The existential doubts that rack you so,
Have whet and honed your mind, till like a sword
You cut through falsehood, let no un-truth go,
And weld your meaning to each well-picked word.
Intensity becomes you, like your eyes,
Dark caves of brown, wherein your wisdom lies.
So rare a jewel, I value you so high,
That all the wealth of Rome I would pass by
If you would to my life true meaning lend,
And hold me in your heart as "Your dear Friend".

26.03.98

# Value

To re-assess our value and our worth,
"We need some time apart", I heard you say,
But did not comprehend the full import,
Nor how your words described love's slow decay.

I thought that I was far beyond compare,
Was free to live, to joke, to laugh, to sport,
And did not need to justify my share
Of this unpleasant moment's bloody birth.

But time has shown that we who – made of clay –
Are frail and fragile, doomed to pass away,
Cannot, in this form, gain any peace
Nor from our pain find solace or release.

So know, full well, that we must all evolve,
And only by this method, all woes solve!

26.03.98

# Musings in the Museum

A pile of grey. A pile of grey. Two heaps displayed:
One close; one placed behind – astray!
Two heaps of grey-white flakes arrayed together,
Soft powdered stones, which through me sent a shiver.
On stepping near to read, we find they're bones,
"One adult and one child" the placard read.
This heap of bones was once a young child's foot!
Grey flakes of stone with ridges, where the marrow
Once gave a sense of life to this still form,
Three thousand years ago when he was born
(And we were far removed from present sorrow).
I mused, unto myself, in Fancy's mood,
"Could this strange pile of bones have once been us,
Have been the Past, of what we've now become?
Did what we are, once manifest as this?
Were we 'adult' and 'child', now long since dead,
Or is this just a nonsense in my head?"

27.03.98

# Desire

*(On Seeing Paco at New Street Station after the interval of some time.)*

To make you part of me was my desire,
To hold you close, to own you, to consume,
To 'somehow' trap the radiance of your bloom,
Enjoy your youth, your beauty and your fire.

Your stunning looks, the brightness of your eyes,
Those potent lips of red, that luring smile
That spreads seductive, innocent of guile,
Yet hides within a tongue, which can chastise.

Those arms of chestnut brown, soft-kissed with hair,
Whose muscular appeal is understated,
For on the court, with racket swing created,
Startle and amaze with brilliance there.

Those thighs, whose strength, agility and speed,
All seem so dormant as you linger there.
And as I watch, a moment unaware,
I feel within the surge of lust and greed.

Thus stimulated, *I* become Desire!
Consumed with fury to attain your form,
To make you mine, and still my inner storm.
But you are not for sale, and I, no buyer.

My heart feels pain for what it cannot own,
For what I had, was clearly just on loan,
And though I knew, I cannot make this known,
That you are you, and made for you, alone.

29.03.98

## Thoughts of you

Throughout the day you slip into my mind
And warm me with the image of your smile,
Then I become ecstatic for a while,
Through knowing you, and calling you my friend.
A joy floods through me while I thus rejoice,
And think of you, your qualities, your gifts,
And all that makes you, you, then my heart lifts,
As I perceive more virtues than my voice
Can ever hope to tell in simple lines;
And so I pass my time in joyful mood,
Relishing your memory like wine,
And making toasts to your eternal good.
  So know my friend, though you are far away,
  You're in my heart, and here you'll always stay.

29.03.98

# Dark

Now comes dark Anger on wild Fury's wings,
Charging in, blood-blinded, doing harm.
Swiftly dealt the death-blow by the arm,
That wields a Jealous sword that cuts and stings.

Now tainted by the ego-thoughts of 'me',
Which all Compassion, Hope and Comfort, kills,
The sea of chopped emotions boils and spills,
Flooding over plains once calm and free.

Turbulent and thund'rous all my dreams,
Now we are set for battle, far apart,
And all is cold sharp steel within my heart,
For nothing now is ever what it seems.

Though you were once my love, that feeling's fled,
And I, in this wild rage, would strike you dead!

07.04.98

# Jealous

I'm Jealous of your boyfriend. Sure, I am!
He has what I desire; what I once had!
No wonder losing you has made me sad,
And leaves me cursing him with F*** and Damn!

Your love was such a wonder, such a gem,
Such a special feeling, none could add,
Or take away, the joy that made me glad,
To be with you, and relish your sweet balm.

Oh how can you prefer this man to me?
D'you think he values you, the way I do?
And can he give you love, the way I could?
You wonder why I'm jealous? Can't you see?
My life is empty, meaningless and crude,
For I have nothing, now that he has you.

07.04.98

# Mess

I've made a mess of all my life's desires,
I've failed you in the greatest truth of all,
My eyes are big, my courage far too small,
And Jealousy now burns in raging fires.
My wild untamed emotions rushing free
Have been the bane of all I've tried to do,
Undone our friendship, turned my hate on you,
And blinded me to what I tried to see.
I am unworthy of your gentle thoughts,
Unworthy of your kindness, love and care.
I can not be objective with my life
But tie you in my complex, heart-string's knots.
I have no love to give, no words to share,
But only drag our Friendship into strife.

07.04.98

## Going for Refuge

*(On hearing the news of his considering withdrawing his Ordination Request.)*

You said your heart no longer yearns to be
An Order Member, or one who's Ordained,
Yet when you talked, your eager mind maintained
An attitude so plain for all to see
That your one true direction could but be,
To manifest the richness you contain,
Which only Going For Refuge would sustain,
And thereby give your life reality.

You say you've lost your way, are all at sea
And have no true perspective, can't go on:
But when you speak, your mind delights my ear,
Your tongue pours forth pure Dharma, till I hear
Your simple words give forth a joyous song,
That cries, "I am a Buddhist! Ordain me!"

08.04.98

# Lost Love

When you loved me, I tasted of Perfection,
Drank in its deep, rich waters to my fill;
Had joy and riches – taste their flavour still.
But you no longer find your satisfaction
In loving me. Instead, in strong reaction,
You have, by your behaviour, helped to kill
The love I had, the love I knew, until
Some other lover offered you distraction.

Oh how I mourn the loss of your sweet love,
The joy I knew, the happiness you brought.
My life was rich, and had a sense of purpose,
A goal to head to, reason to improve.
But now it lacks all substance, is of naught,
All joy is gone and I've become superfluous!

08.04.98

## Paco

I told my love I loved him so,
He cursed me for a fool.
I shed a tear, and wondered how
My love could be so cruel?

But I was wrong, I had no love,
For he did not love me.
So now I hate the one I loved,
And curse eternally.

08.04.98

## Love Lost

To bask within the sunshine of your love,
To be again with you as once we were,
To taste the joy fresh love must always bring,
And in that pleasure rouse our hearts to sing
of rapture, and of bliss, and of the move
from strong dissatisfaction – our first spur –
To calm, serene, contentment. Rapturous state!

Oh would those heady days return once more
And bless us with the wealth of their estate,
To find in each the treasure we adore,
And taste the sense of wholeness we once had.
I'd pledge my love would never be undone,
And sell my soul to keep the Devil glad,
If only I could bask in your love's sun.

12.04.98

## The death of love

*(After speaking to Paco on the phone about his split from his boyfriend.)*

Unceasing is that ache within your being,
Until you think, "No future can I have.
This world can have no meaning from now on!"
There seems no way, no hope, of ever freeing
Your body and your mind from shattered love.
The world seems bleak, and nothing can be done.
But pause, my friend, delay your darkest thought.
Although your pain is great, it will not last,
For even mountains crumble into dust,
And all your hurt will one day turn to naught,
As from your heart this current love is cast
And Passion is supplanted by fresh lust.
   So bear your pain, and know that it will pass,
   And do not be a martyr to this cross.

15.04.98

# Alone

Because we do not meet as once before,
My life is poorer and my heart is sad.
Gone are the foolish pranks that made me glad,
Gone the silly jokes – our private lore -
As if New York had somehow closed a door
And we no longer owned what once we had:
As if the love I feel is viewed as 'bad',
And meeting me, for you, is now a chore.

Forgive me, friend, you're stronger far than I.
I do not have the clarity of thought,
Nor all the sharp intensity you own;
So, friendless now, I soldier on alone,
And feel, without you, my life is but naught,
For on your love and friendship I rely.

18.04.98

# Kalyana Mitrata

Our Love's resolved to Duty, me to thee,
'Tis not a chore, but task I gladly do,
To help you in pursuit of something True.
Thus I accept responsibility;
Acknowledge that your path will not be straight,
And hope that I can give some good advice,
And steer you from the pathways fraught with vice,
Help to make Compassion from your Hate,
And always be a Friend, when friendship's sought.
I also know our task is not one-sided,
And though you learn, I also can be taught,
For your rich mind much Wisdom has provided.
   So let's be friends, rejoice in one another,
   And learn to Love, although we are not lovers.

19.04.98

## Mutual

Think not, I do not comprehend your pain,
Or suffer with you, as your tale relate;
Your words arrive like echoes in my brain,
Revive again old wounds of love and hate.
To feel so deeply that intense pure love;
To want to share and feel reciprocated;
To wear that tempest like a passive slave:
In all this pain I too participated:
Those tearful moments when the heart must break,
The joys remembered at each tender act,
That kindly gift; that now much altered 'take',
Which threw a whole new meaning on some fact.
　Yes, love is full of turmoil, pain and doubt,
　And being IN love is best when you get OUT!

20.04.98

# Grief (Mourning for Lost Love)

There is a grief, a pang, an ache so deep,
That tears cannot relieve its buried woe,
So all-embracing, void of outward show,
That none can know, or hope to make the leap,
From such discomfort - eased alone through sleep
Into the terror of a death so slow,
Which creeps, as if still trapped so far below,
Where roots and all its history will keep
It hidden from mankind for ever more.
But this can never be, for love must die,
And sorrow unexpressed will bubble o'er
And spill its salt-filled waters on the flesh
That, careful till that time to hide the lie,
Is pleased at last this anguish to unleash.

24.04.98

## Spanish Tart

*(A Fragment.)*

You think me mad! I see it in your eyes!
Come. Come. Put off that feigned disguise!
You think me mad and there we have it!
Well, listen well. No! Please! Stay! Sit!
I'll tell a tale will make your face aghast,
Terrify your heart from first to last.
No. Do. Please. Sit. You'll find me kind,
And not some crazy madman out of mind.
Please! I beg you Sir. Take up some wine.
Would someone so insane drink wine so fine?
And would he be so eloquent and clear,
And able to placate irrational fear?
See, you're thinking now. Have paused for thought,
My momentary outburst's been forgot.
There! Good! Relax. I've won you now.
Your eyes show curiosity, your brow's
Contorted like a trout caught on a hook.
So settle back. I'll tell it like a book,
And turn the pages slowly so you may
Be lured along, enraptured by my tale,

Yet horrified to understand the scale,
The vast extent, and scope of all my pain,
That rages like a torrent through my brain.
And then at last you'll know my tortured heart.
Feel the terror, jump to take my part,
And cry with me, run howling to the skies,
And know what sets this 'madness' in my eyes.

Two years ago.
Before you had the knowledge of my name.
(Before my crazy sorrow, and my shame.)
I met a boy. A youth. A fit young man!
And with that simple fact my woe began.
His eyes were brown, and dark, and deep, and full,
And drew me to them, exercised a pull,
As if the Night-time sky had been distilled,
And set within a Marble, black and still:
Or ran like blood, the brown of swollen streams,
Made mad by Rain's torrential, downward teem.
Flashing and sparkling, dark, and yet so light,
So rare, so perfect, such a luring sight.
Huh.
You smile! And write me off like some foul scourge.
"A fool in love, o'ercome by lust's wild surge."

Well. Sniff your snuff and think yourself content,
I *was* in Love, But that's not what I meant.
His eyes were where the magic of his soul
Worked out its potent witchcraft, which then stole
Into the very centre of my brain,
Corrupting me in each and every vein,
Until I had no choice but follow him,
Become his slave, submissively give in.
There was no other option, no clear way.
Believe me, sir. Take heed of what I say.
This is no jest, no lightly gathered tale,
This is the Truth. The facts. The whole detail.
Oh yes. I see again that inward laugh,
That jolly jape, with Evening's ale to quaff,
A lively fabrication for your friends,
Regale them with dementia, and its ends.
Oh yes.
It sounds like pretty poetry, and farce.
A woman's tale. Some heavenly romance.
A love song, sounding sweet, and gently sung,
So thickly sugared, and so clumsily begun,
Intent to lure new lovers to embrace,
And heal their fleeting strife, with one warm kiss,
But no.

That is not what I meant to say at all.
It does not match that motive. Not at all!
Nor never can be, though it may
Recall the outward circumstance, the way,
The mood, the unassuming first delight,
That gently swept upon me at his sight.
Those eyes, I see them yet. Those eyes,
Black, as black, as black, as jet black skies.
Yet full of life, and light, and wild surprise,
And sparkling, bubbling o'er with wild surmise,
As if he too could somehow sense the pulse,
The surge of wild emotion, like a curse,
The flowed between our two enchanted forms.
Believe me, sir. I've sailed upon Love's storms,
Been shipwrecked on the rocks of Jealous hate,
Been racked with strong Desires, and tempted fate,
Been wrapped within Dependence, larva-like,
Never to emerge and take to flight.
Chained to strong Attachments, sorely pained,
And wept with grief, more Pain than Pleasure gained.
So do not think Love's never seen my face,
Or used me badly, spurned me from her grace,
But this was different. This was something new,
Some flavour from some strange, exotic brew.

I went to him. You look surprised!
Why should I not? I'd gazed into his eyes.
He knew my mind. He knew my base desire.
And kindled with his smile a raging fire,
Which, if his eyes alone had drawn me in,
That smile now bound me totally to him.
You smirk again, as if this was a jest,
Some party trick, brought out to prove the zest,
Add sparkle, to a dull and sordid dish,
By making it more spicy than you wish.
Believe me, sir, each word I speak is true,
You cannot comprehend just what I do,
Or why I need to make you understand,
The Truth I speak, the sorrow I command.
I said I went towards him, and I did,
His eyes behind long lashes, flashing, hid.
He fluttered them as he began to speak,
With words of kindness, smooth, and silky sleek,
That I, enthralled, heard music in my ears
And marvelled at the wisdom of his years.
His spell was cast. The final web was spun,
And I, trussed up and helpless, was undone.
He spoke of life, of poetry, and love,
He told of how his heart was drawn above,

Taken to some place beyond this earth,
Where he would live again, some bright new birth,
And how he only needed one true friend,
To make this dream come true, this longing end.
It was a chance too good for me to miss,
I offered up my friendship with a kiss,
And he accepted, melted in my arms,
And swole my joy, with all his manly charms.
He came to stay. We shared all that I had,
For he had nothing, being just a lad.
At first my joy was boundless like the sky,
There was no gap between us, He or I.
We mingled so completely, mind to mind,
We shared all pleasure, our interests intertwined.
I never tasted love of such a depth,
Or had another offer so much warmth.
He made my life complete, filled every need,
And stilled the hollow grasping of my Greed.
No. Please. Take more. Fill up your glass.
I do not drink, taste nothing now, alas.

29.04.98

## The Scent of Friendship

*(After a remark made on the telephone by Paco.)*

The scent of Friendship lingered in the room,
Long after you had left me here alone:
A sweetness that your kindness could assume,
And spread around, not hoarding as your own.
All day I caught the fragrance of your being,
Perfuming my remembrance with your air,
And in my mind, your smile, I kept on seeing,
Aware, of course, that you were nowhere near.
But like the incense smoke, that rises straight,
And permeates the air on which it floats,
So too, does Friendship slowly saturate,
And scents the lives of those on whom she dotes.
  Thus Friendship is a fragrance strong and sweet,
  Whose perfume adds to life a scent complete.

08.06.98

# Wuthering Heights

*(A critique.)*

Oh do not think I meant to mock or scorn,
My views were simply views I held in truth,
And not from any form of malice born,
Nor any way a comment on your youth.
For you, though young, are twice as wise as I,
Your mind is clear and sharp, your thoughts are bright,
And I would be a fool to even try,
Through mean endeavour your good name to slight.
For if I ever loved, I loved but you,
And all my life seems geared towards the time,
When I would realise that this was true,
And celebrate this fact in verse and rhyme.
   So no offence to you was ever meant,
   But if you took some, I, at once, repent.

25.08.98

# Sadomasochism

I hate you when you cast me off alone,
Reject me and desert me in this place.

*(And yet this haughty action I condone,
Enjoy these guilty feelings of disgrace.)*

You stood me up, went off with someone new,
Made me look foolish, sheepish and forlorn.

*(Yet I feel good when being abused by you,
Enjoy my anguish, revel in your scorn.)*

This sadomasochistic trait in me
Seems so outrageous, so against the norm,

*(And yet it is the way I have to be,
The way I am, the way that I was born.)*

My love is strange, I know it. I'm perverse.
But, compared to yours, is it really worse?

25.08.98

## Betrayal – Pizza Express after the film

That hell could be so deep and full of pain,
I never knew till I saw your sweet face,
Contorted and distorted by the strain,
Of knowing that a breach had taken place.
The man in you had shrunk to fit the boy,
Whose brown and injured eyes conveyed your grief,
Those pools of living liquid, void of joy,
That struggled smile that offered no relief.
Never had I felt another's pain
So sharply as I felt the pain in you.
Here before my eyes I watched life drain,
And saw betrayal cut your heart in two.
  In that one instant, Innocence was gone,
  And Knowledge and Experience were born.

25.08.98

## Paco, on Love – for Stephen.

As if our time on earth was not enough,
A curse is added to our suffering,
(A curse that makes our smoothest passage rough,
A curse that makes the mad with madness sing.)
The Curse of love: of bitter-sweet Romance;
Of Longing and Desire wrapt up in Lust,
Of Death, of Greed, of Hatred all at once,
Grinding the heart and mind and loins to dust.
Oh what an evil mind created this?
This torment of the flesh and of the heart,
One moment I am bathed in Heav'nly bliss,
The next by Hellish fury torn apart.
   I know the Truth, I know the way to be,
   That only by renouncing am I free.

25.08.98

# The Leaf

Within my palm I hold the fragile leaf,
Nay skeleton, mere shadow of itself.
And as I stare in wonder, feel its grief,
I realise the frailty of my self.
I realise that you and I must die,
I realise that you are yet so young,
I realise the tears you've yet to cry,
I realise your journey's just begun.
Oh sweet and gentle youth – Love of my life!
What sorrow must your young heart yet endure,
What pain and anguish, heartache, terror, strife,
And all this heap of woe, yet to mature?
Oh would that I could be some stalwart oak,
Some solid trunk to shoulder all your woes,
A heart so large all sorrows it would cloak,
And shelter you until your anguish goes.
  Then like a leaf you'd float towards the earth,
  And in that moment know my friendship's worth.

25.08.98

## Insensitive

My mental equal, and in ways my sire,
I thought of you – though young – as my true peer;
Resilient and robust – an ancient lyre –
From which no tune could pluck a feeble tear.
But manly though you are, you're still a boy,
A youth who seeks frivolity and joy,
Who needs protection, shelter, comfort, care,
Who needs a loving heart his heart to share.
But I was much too crude this fact to see,
Too blind to your emotions or your need.
I treated you as if you thought like me,
And were a brother glutton in my greed.
  But seeing you feeling hurt has hurt me more,
  Released my tears, and cut me to the core.

25.08.98

## Another of the same

I never saw such waves of pain before
As those that crossed your face on that dark night.
I watched that hurt destroy your very core,
And leave you lonely, empty, pale with fright.
We knew in that dire moment, one dark truth:
Our love had died, and we were henceforth two.
Though young, you were no more a naïve youth,
For Knowledge had matured a part of you.
I lost you there and then. We fell apart.
Acknowledged that we were no longer one,
And though strong feelings lingered in each heart,
We knew romantic love had been undone.
  Thus seeing your poor heart break, has broke mine,
  My world is dark, for you no longer shine.

25.08.98

# Choice

Life is a choice that I must surely make,
On one hand stands my lover, bold and strong,
The other holds a Buddha, wide awake!
Who's drawing me from out the madding throng.
I want them both, my heart goes out to each,
Each has its charms, its lure, its crock of gold.
Each has much merit, and they both can teach,
And lessons from the book of Life unfold.
Oh would that I could somehow own them both,
Be blessed with two in one, not one or t'other,
My mind when forced to choose is full of sloth,
And I, confused, find choosing too much bother.
  How strong the pull of mundane life can seem,
  Yet Sunyata is such an EMPTY dream!

25.08.98

# Ode to Paco

*(A Fragment.)*

Tonight my idle pen awakes
And coaxes, teases, pulls and shakes
The red ink from its slumbers long,
To shape the outlines of my song.
In these few rudimentary words,
That cannot plead my heart to you,
Nor write the love I feel's your due,
I will, with humble hopes and dreams,
Share the joy your presence brings.
I'll tell how much you mean to me,
How precious, rich and rare you be.
How, when I hear your lively voice,
My heart leaps up, my ears rejoice.
And when your image, through my brain,
Breezes in, then out again,
It feels as if I've just been kissed,
Some being divine has shed his bliss,
And wrapped me in his mantle sweet,
To taste delight, made me complete.
Oh shaking pen! Oh ragged scrawl!

Release the feelings long held thrall,
Within this tyrant heart of mine.
Those feelings of delight and joy,
Those feelings, which my fears destroy.
Feelings which I would express,
But feelings which I must suppress.
Those feelings, like that beast of yore,
That hundred-headed Typhon's roar,
Whom Zeus, with flaming thunderbolt,
Subdued and quelled with wild revolt,
And buried deep within the rock,
Below old Aetna's smoking stack.
And, like that monster lying there,
My feelings smoke and rumble still,
My unexpressed emotion boils,
My Anger thus my Love embroils.
Convention has me in Her grip,
Castrates me with Her Moral whip,
And drains my feelings dry.
I cannot wrap my arm around
Some innocent and happy child.
I cannot kiss, or show affection,
Share my joy at Life's perfection.
I cannot play with innocence,

Or be a friend with joyful zest,
For guilt and pain, and fearful thoughts,
Cloud my vision, curb my lot,
Until I cannot say out loud,
"I love you!" in a voice so proud.
So though I can be light and gay,
I am unwilling to be Gay...

23.02.99

# (New York) – Drowning

I could not reach you.

I stood on Brooklyn Bridge
Holding out my hand
But you were already
Five fathoms drowned.

I could not save you
Never give you love
My craving was a wave upon a shore
That never would arrive.

In Central Park you lectured me
Told me not to 'hold'
Berated me with how I was attached.
I clung to your words like a drowning man
But could not understand.

In the Wall Street Sauna
I watched you soar
Up above the image of your dream

Arm in arm with the idol of your screen
But who was I?

You took his home phone-number
Talked of days in the country
You'd spend the week with him!
But who was I?
You stood on Brooklyn Bridge
Watching while I drowned.

We were fighting for survival through our hurt.
I did not understand.
I wanted you, and wanted you for me.
You wanted him,
I could not understand.

You threw a little thing off Brooklyn Bridge.
I tried to help you struggle with the pain
I heaved I pushed
I did not understand.
The little thing you threw was me.

29.08.99

# Motorway

Tears in a motorway service station
Burn down your cheeks and gouge your flesh,
Lancing my heart, they tear my tissue,
Shocking complacency out of my face.

“Please don’t cry, I just can’t bear it!“
Silent words in my speechless brain.
“Please don’t cry, I just can’t bear it!”
Eyes ablaze in the smirring rain.

All your youth and childhood troubles,
All your future deaths were there.
All your love, and my cruel coldness,
Burned, and bruised, and burbled there.

What can I say? You chose another!
What can I say? I love you still!
So much hurt, I could not bear it,
So much hurt in one salt tear.

Please don't cry, I can't endure it.
Please don't cry, I'm not to blame.
Please don't cry, I love you truly.
Please don't cry, just say goodbye.

Tears in a motorway service station
Spoke of love and life and death.
Tears in a motorway service station
When you released your heavy grief!

28.08.99

## AUTOPISTA

Lágrimas en una estación de servicio
queman tus mejillas y resquebrajan tu piel.
Lanceando mi corazón, sacuden tu cuerpo
y arrancan de mi rostro la complacencia.

"No llores por favor, que no puedo soportarlo"
Palabras de silencio atrapadas en mi mente.
"No llores por favor, que no puedo soportarlo"
Ojos candentes y cuatro cristales empañados.

Todos los conflictos de tu adolescencia,
todas tus futuras muertes, estaban ahí.
Todo tu amor, y mi cruel frialdad,
abrasaban, ardían, hervían ahí.

¿Qué puedo decir? ¡Escogiste a otro!
¿Que qué puedo decir? ¡Te sigo queriendo!
Tanto dolor, no podía soportarlo,
tanto dolor en una lágrima salada.

No llores por favor, que no puedo soportarlo.
No llores por favor, no hay nadie a quien culpar.
No llores por favor, que te quiero de verdad.
No llores por favor, despídete ya de mí.

Lágrimas en una estación de servicio,
hablaban del amor y de la vida y de la muerte.
Lágrimas en una estación de servicio,
alivio de tu pesar.

(*Motorway* by Kovida. Translated by Francisco Domínguez Montero.)

Mayajala is a word from the ancient Indian language of Sanskrit which means 'web', or 'net' of illusion.

For more information about Mayajala Books please visit our web site at:-

www.mayajala.org